# Only Oracles: Solo Journaling

*Tools for Solo Role-Playing*

*Adventures*

*Only Oracles: Solo Journaling v1.0: August, 2024*
*Copyright©2024, Basunat*
*Art: JAR*

*For more Solo RPG books, games, settings and adventures, visit:*

*Basunat Drivethrurpg and Basunat itch.io*

Basunat

2024

# TABLE OF CONTENTS

In the quiet solitude of my room, illuminated more by the flickering glow of a desk lamp than by the warmth of the sun, I found myself submerged in worlds crafted by my own imagination. A solitary player among the many voices of characters I created, I began to realize something unsettling. The oracles I relied upon—the magic tables, the vague snippets of advice, the fragmented prompts—they felt insufficient. They lacked the nuance, the vibrancy, the pulse of the stories I longed to tell.

When I first embraced the realm of solo RPGs, it was with exuberance. Here, I could be the architect of my narratives, unfettered by the constraints of a Dungeon Master or a group. Yet, exploring these sprawling adventures, the trappings of traditional oracles often felt like a weight. They were clunky tools, rigid frameworks that didn't bend to the wild shape of my imagination. I wanted—no, needed—something more profound, something that resonated with the heart of the tales I wished to weave.

The dissatisfaction simmered within me, a nagging itch whispered by the spirits of my stories. I would sit with my notebook, scribbling furiously, the ink etching my thoughts onto the page like a pulse of vitality yearning to escape. Through the rhythms of my solo journaling, a vision emerged—a desire to curate oracles that rather than constraining, would liberate. They would not be mere tables

of options; they would sing with specificity, each oracle steeped in the flavor of distinct genres, each result a thread woven into the fabric of my unfolding adventures.

And so, I began a journey—a pilgrimage through the expansive landscapes of narrative genres, pouring myself into the exploration of their depths. Fantasy, science fiction, horror—the boundaries blurred as I began to see the potential for oracles that truly captured the essence of these realms. No longer would they be generic responses, but mirrors reflecting the emotional landscapes I traversed. I yearned for a connection, a dialogue between the oracle and my story, not just a simple transaction of prompts and random choices.

Each genre's oracles came alive, breathing in the air of their respective narratives. In the fantasy section, the oracles whispered secrets of missions and quests, revealing encounters that felt electric, infused with the tension of unspoken words. In the realm of horror, they clawed at the edges of darkness, unveiling dread and suspense, demanding my imagination delve deeper into the abyss. The oracles became companions on this solitary journey—empathetic guides navigating the labyrinth of my creativity.

My notes were scattered, sprawling observations inked in the late hours when the world hushed and my mind buzzed with possibility. They were unrefined, visceral eruptions of inspiration. But they held a promise, a possibility waiting to be unlocked. Each jotted line, each scribbled oracle whispered tales of adventure yet to be lived, of encounters yet to be forged. They were vessels of my solitude—a humble testament to the beauty of one's own voice in a universe brimming with stories.

As I began to type and organize these reflections with the help of new tools, weaving together the myriad oracles birthed from my frustration and my revelations, I felt a sense of purpose solidify. This book would be an offering to others who, like me, might grapple with the limitations of traditional oracles, enticing them to explore the textured landscapes of their own imaginations. It would be a beacon for solitary adventurers, each oracle a key to unlock not just outcomes, but deeper connections to the worlds we create.

In crafting these oracles, I realized that they are, in essence, intertwined with my own psyche. They are projections of desires, fears, and dreams—a mirror held up to the player who seeks not just a diversion, but a reflection of their inner landscape. No longer am I confined to the mundane; I wanted each encounter to evoke emotion, to resonate with the core of what makes us human—the joys and sorrows, the triumphs and failures, the intense yearning for connection.

As I pen these words, I am reminded of the intricacies of poetry—a raw exploration of the human condition layered with delicate imagery. My writing, too, aims to capture those fleeting moments of clarity between the chaos. The oracles I crafted echo human sentiments, touching the ache of isolation while simultaneously celebrating the richness of individual experience. They are lifelines cast into the ocean of solitude, helping players navigate the tumultuous waters of imagination alone yet profoundly connected.

This first book, the beginning of a series of genre-specific volumes of oracles, was born from my journaling solo sessions. It is a love letter to the art of storytelling, the culmination of countless hours spent exploring the brilliant limits of creativity. As you turn the pages, letting the oracles guide you, you may find the essence of your own tales blossoming, intricate and beautiful, in response to the clues offered by these carefully selected suggestions. Believe me, there are myriad paths within you.

And so, travel with me on this quest, let us delve deep into the realms of possibility, unearthing the treasures that lay dormant, waiting for the whisper of an oracle to awaken them. Accept the invitation to dance in the rhythms of your imagination, to ignite the spark of creativity that resides within, and to embrace the tales you have yet to tell.

Welcome to the journey of crafting your stories, enriched by the power of your memories and your imagination, and steeped in the nuances that give life to every story you are able to create.

# Oracles and Solo Journaling RPGs

*Solo Journaling RPGs focus on introspective and personal storytelling, where players record the experiences and thoughts of their characters. This book offers oracles that inspire deep reflections, emotional developments, and meaningful decisions. From personal revelations and significant encounters to internal challenges and moments of growth, each oracle is designed to enrich personal narratives and create a journal full of life and detail.*

### Oracle Categories

1. Personal Revelations: Moments of Insight
2. Significant Encounters: Influential Meetings
3. Internal Challenges: Personal Struggles
4. Moments of Growth: Development and Change
5. Emotional Turmoil: Inner Conflict
6. Important Decisions: Choices That Matter
7. Secrets Unveiled: Hidden Truths
8. Dreams and Visions: Subconscious Messages
9. Personal Revelations: Discovering Inner Truths
10. Conflicting Desires: Internal Conflicts
11. Epiphanies: Moments of Clarity
12. Legacy and Memory: Reflections on the Past

*In the quiet corners of libraries, forgotten manuscripts reveal the ancient art of solo journaling. Sages whispered that these journals held portals to parallel worlds—a place where ink and imagination merged. The Oracle Keepers, reclusive scholars, safeguarded these secrets.*

*The first Oracle, a hermit named Siwa, inscribed her musings on parchment. As she wrote, her words transcended paper, pulling her into vivid realms. She chronicled her inner battles, love affairs, and quests for self-discovery. Her journal became a living artifact, echoing across time.*

*Siwa's legacy inspired others. Knights, poets, and dreamers took up their quills, seeking solace in solitude. They recorded their triumphs and heartaches, their whispered confessions and silent screams. The Oracle Keepers nodded in approval, for each entry wove a new thread in the cosmic tapestry.*

*And so, the tradition continued—an intimate dance between ink and soul. The Oracle Keepers vanished, leaving behind their journals as beacons. Those who stumbled upon them discovered a refuge—a place where they could be heroes, villains, and everything in between.*

*Remember that your journal is more than parchment and ink. It is a gateway to worlds unseen, a mirror reflecting your essence. Write boldly, for the oracles await your touch.*

# What Are Oracles?

*S*olo journaling RPGs offer a unique and deeply personal form of storytelling. At the heart of these games lies the concept of Oracles, which are random prompts or tables used to inspire creativity and guide the narrative.

Oracles are tools that generate ideas, situations, or questions, acting as catalysts for the player's imagination. They help overcome writer's block, introduce unexpected twists, and add depth to the story. Oracles can take many forms, such as lists, charts, or cards, and they cover various aspects of the narrative, including plot developments, character traits, and environmental details.

### The Role of Oracles in Solo Journaling

*Inspiration:* Oracles provide a spark of inspiration, offering new directions and possibilities for the story. They encourage players to think creatively and explore different narrative paths.

*Guidance:* By offering structured prompts, oracles help maintain momentum and focus in the storytelling process. They can guide the player through complex plot points and character developments.

*Unpredictability:* Oracles introduce an element of randomness, making the story more dynamic and engaging. This unpredictability can lead to surprising discoveries and emotional moments.

*Depth:* Oracles encourage players to delve deeper into their characters' psyches, exploring their motivations, fears, and desires. This leads to richer, more nuanced storytelling.

### The Beauty of Solo Journaling

Solo Journaling RPGs offer a meditative and introspective approach to storytelling. They allow players to explore their inner worlds, confront personal challenges, and experience growth through their characters. The act of journaling itself can be therapeutic, providing a creative outlet and a means of self-expression.

In solo journaling, the journey is just as important as the destination. The process of uncovering your character's story, one oracle at a time, creates a rich tapestry of experiences that is uniquely your own. Solo Journaling RPGs offer a powerful and rewarding way to tell your story.

With these oracles and insights, you have the tools to begin a deeply personal and creative storytelling journey.

*Let yourself be guided by the clues of the oracles, but always remember that the heart of the story lies in your own imagination and the unique path you carve for your character.*

## How to use this oracles

To use oracles effectively, follow these steps.

1. **Select an Oracle:** Choose an oracle that fits the current context of your story. It could be related to character development, plot twists, or introspective moments.

2. **Roll or Draw:** Roll a die or draw a card to generate a random prompt from the oracle.

3. **Reflect and Write:** Take a moment to reflect on the clue and how it fits into your story. Write down your character's thoughts, reactions, and actions in response to the prompt.

4. **Integrate and Continue:** Integrate the new development into your narrative, allowing it to influence the direction of the story. Continue writing, using additional oracles as needed.

# 1. Personal Revelations: Moments of Insight

*As I sat before my journal, quill in hand, I felt the weight of countless stories yet untold. The blank page stared back at me, a canvas of infinite possibility. It was then that I remembered the whispered tales of the Oracle Keepers, and the power they held to unlock the doors of imagination. With trembling fingers, I reached for the ancient tomes, each one a gateway to a different facet of my character's soul.*

*The first oracle I discovered, Personal Revelations: Moments of Insight, became my beacon in times of introspection. When my character stood at a crossroads, unsure of their path, I would turn to this oracle. Its whispers would reveal hidden truths about their past, illuminating the way forward. In moments of quiet reflection, by candlelight or beneath the stars, this oracle became my confidant, helping me uncover layers of my character's psyche I never knew existed.*

# 1. Personal Revelations: Moments of Insight Oracle

1. Realizing a hidden talent.
2. Understanding a past mistake.
3. Recognizing an old fear.
4. Discovering a newfound confidence.
5. Embracing a personal truth.
6. Uncovering a suppressed memory.
7. Accepting a difficult reality.
8. Noticing a pattern in behavior.
9. Finding inner peace.
10. Acknowledging an unspoken desire.
11. Confronting a long-held belief.
12. Realizing the importance of forgiveness.
13. Feeling a deep connection to nature.
14. Understanding the value of solitude.
15. Recognizing the influence of a mentor.
16. Discovering a passion for a new hobby.
17. Embracing vulnerability.
18. Finding strength in resilience.
19. Accepting the inevitability of change.
20. Realizing the impact of a past decision.
21. Feeling empathy for a former rival.
22. Understanding the weight of responsibility.
23. Recognizing the need for self-care.
24. Embracing a long-forgotten dream.
25. Finding beauty in imperfection.
26. Understanding the power of choice.
27. Confronting an internal bias.
28. Realizing the value of patience.
29. Accepting one's own mortality.
30. Recognizing the need for balance.
31. Understanding the significance of a ritual.
32. Finding solace in a tradition.
33. Embracing a cultural heritage.
34. Recognizing the strength in community.
35. Understanding the power of silence.
36. Discovering the joy in helping others.
37. Realizing the importance of boundaries.
38. Accepting the past without regret.
39. Embracing the present moment.
40. Recognizing the promise of the future.
41. Understanding the nature of forgiveness.
42. Finding clarity in confusion.
43. Embracing the unknown.
44. Understanding the role of fate.
45. Recognizing the power of intention.
46. Accepting personal limitations.
47. Finding inspiration in adversity.
48. Understanding the importance of hope.
49. Recognizing the influence of the environment.
50. Embracing personal growth.
51. Realizing the interconnectedness of all things.
52. Accepting the unpredictability of life.
53. Understanding the impact of words.
54. Recognizing the strength in humility.
55. Embracing the power of creativity.
56. Finding joy in simplicity.
57. Understanding the significance of a promise.
58. Recognizing the value of perseverance.
59. Accepting the gift of time.
60. Understanding the role of intuition.
61. Embracing the spirit of adventure.
62. Finding wisdom in experience.
63. Recognizing the importance of rest.
64. Accepting the necessity of change.
65. Understanding the power of dreams.
66. Realizing the value of reflection.
67. Embracing the journey of self-discovery.
68. Understanding the impact of love.
69. Recognizing the importance of faith.
70. Accepting the challenge of self-improvement.
71. Finding peace in letting go.
72. Understanding the nature of sacrifice.
73. Embracing the concept of destiny.
74. Recognizing the role of forgiveness.
75. Accepting the beauty of uncertainty.
76. Finding strength in diversity.
77. Understanding the significance of trust.
78. Recognizing the power of empathy.
79. Embracing the journey of healing.
80. Understanding the impact of kindness.
81. Recognizing the importance of gratitude.
82. Accepting the lessons of failure.
83. Finding joy in accomplishment.
84. Understanding the role of courage.
85. Recognizing the value of loyalty.
86. Embracing the concept of harmony.
87. Understanding the importance of mindfulness.
88. Recognizing the strength in compassion.
89. Accepting the gift of forgiveness.
90. Finding clarity in chaos.
91. Understanding the nature of truth.
92. Recognizing the power of resilience.
93. Embracing the beauty of impermanence.
94. Understanding the impact of generosity.
95. Recognizing the importance of self-expression.
96. Accepting the necessity of growth.
97. Finding peace in acceptance.
98. Understanding the value of friendship.
99. Recognizing the significance of legacy.
100. Embracing the wisdom of ancestors

## 2. Significant Encounters: Influential Meetings

*As my journey progressed, I found myself drawn to Significant Encounters: Influential Meetings. This oracle breathed life into the world around my character, populating it with vibrant personalities and potential allies or adversaries. When the solitude of my writing threatened to become overwhelming, I would call upon this oracle to introduce a chance meeting, a mysterious stranger, or a long-lost friend. These encounters added richness to my narrative, creating a tapestry of relationships that shaped my character's destiny.*

## 2. Significant Encounters: Influential Meetings Oracle

1. A wise mentor offers guidance.
2. A childhood friend returns with news.
3. A stranger provides a crucial clue.
4. An old rival seeks reconciliation.
5. A former lover brings closure.
6. A family member shares a hidden secret.
7. An unexpected ally appears in a time of need.
8. A spiritual leader offers a profound insight.
9. A traveler shares a tale of adventure.
10. A respected elder imparts wisdom.
11. A neighbor reveals a shared history.
12. A lost pet is found, bringing joy.
13. A teacher challenges old beliefs.
14. A child asks a thought-provoking question.
15. An artist inspires with their work.
16. A performer moves the heart with a song.
17. A writer shares a life-changing book.
18. A healer offers both physical and emotional care.
19. A philosopher debates a fundamental truth.
20. An inventor introduces a transformative idea.
21. A scientist explains a groundbreaking theory.
22. A historian recounts a forgotten event.
23. A stranger offers unexpected kindness.
24. A gardener reveals the secrets of nature.
25. A chef shares a meal that evokes memories.
26. A soldier recounts stories of bravery.
27. An explorer describes uncharted lands.
28. A diplomat mediates a longstanding conflict.
29. A judge delivers a surprising verdict.
30. A priest performs a meaningful ritual.
31. A dancer expresses emotion through movement.
32. A painter captures a soul in a portrait.
33. A sculptor shapes a vision from stone.
34. A musician composes a haunting melody.
35. An actor portrays a transformative role.
36. A playwright crafts a powerful narrative.
37. A director brings a story to life on stage.
38. A producer creates a memorable experience.
39. A critic offers a thought-provoking perspective.
40. A comedian brings laughter in dark times.
41. An athlete demonstrates perseverance.
42. A coach motivates with unwavering belief.
43. A sailor shares tales of the sea.
44. A pilot describes the thrill of flight.
45. A mountaineer speaks of overcoming peaks.
46. A diver reveals the wonders beneath the waves.
47. A rider recounts adventures on horseback.
48. A driver narrates journeys across lands.
49. A mechanic repairs more than just machines.

50. A tailor weaves stories into garments.
51. A jeweler infuses meaning into their craft.
52. A merchant deals in more than just goods.
53. A librarian unlocks the magic of books.
54. An archivist preserves the voices of the past.
55. A curator assembles a narrative through artifacts.
56. A collector finds joy in the rare and unique.
57. A philanthropist shares wealth with you.
58. A volunteer gives you work or money.
59. A farmer tends to the cycles of life.
60. A fisherman respects the balance of the waters.
61. A forester guards the secrets of the woods.
62. A ranger protects the harmony of the wild.
63. A builder constructs dreams into reality.
64. A smith forges strength and artistry.
65. A mason carves legacy into stone.
66. A carpenter crafts warmth from wood.
67. A potter shapes beauty from clay.
68. A weaver creates tapestry from thread.
69. A healer tends to both body and soul.
70. A midwife witnesses the beginning of life.
71. An undertaker honors the end of life.
72. A detective unravels a mystery.
73. A lawyer advocates for justice.
74. An activist fights for change.
75. A politician and the complexities of power.
76. A monarch embodies the burdens of rule.
77. A spy uncovers secrets with caution.
78. A guard defends with loyalty.
79. A scout explores with curiosity.
80. A hermit seeks solitude and truth.
81. A witch brews wisdom in every potion.
82. A shaman communicates with the spirits.
83. A monk meditates on the path of truth.
84. A nun devotes life to service.
85. An oracle sees the threads of fate.
86. A fortune teller glimpses into the future.
87. A dreamer imagines worlds unseen.
88. A wanderer roams with purpose unknown.
89. A guide leads others to discovery.
90. A chieftain governs with wisdom and strength.
91. An outcast finds community in the fringes.
92. A revolutionary ignites the flames of change.
93. A refugee seeks a place to call home.
94. A pioneer ventures into the unknown.
95. A drifter moves with the currents of life.
96. An artisan imbues objects with soul.
97. A magician performs wonders beyond understanding.
98. A guardian protects sacred knowledge.
99. A seeker quests for truth.
100. A sage imparts the final lesson.

# 3. Internal Challenges: Personal Struggles

*In times of hardship, when my character faced their darkest hours, I turned to* Internal Challenges: Personal Struggles. *This oracle forced me to confront the demons that lurked within my character's heart. It pushed me to explore their fears, their weaknesses, and their most closely guarded secrets. Through tear-stained pages and ink-smudged confessions, this oracle helped me forge a heroine tempered by adversity.*

# 3. Internal Challenges: Personal Struggles Oracle

1. Battling self-doubt.
2. Overcoming fear of failure.
3. Struggling with loneliness.
4. Confronting a guilty conscience.
5. Dealing with anger management.
6. Battling an addiction.
7. Struggling with perfectionism.
8. Managing anxiety.
9. Overcoming a phobia.
10. Dealing with grief.
11. Struggling with jealousy.
12. Confronting self-sabotage.
13. Managing stress.
14. Battling sadness.
15. Overcoming procrastination.
16. Dealing with impostor syndrome.
17. Struggling with indecision.
18. Confronting a fear of success.
19. Battling a negative self-image.
20. Managing chronic pain.
21. Dealing with betrayal.
22. Struggling with trust issues.
23. Confronting existential dread.
24. Battling feelings of inadequacy.
25. Managing work-life balance.
26. Struggling with body image.
27. Confronting a midlife crisis.
28. Battling regret.
29. Dealing with a traumatic memory.
30. Struggling with change.
31. Confronting an identity crisis.
32. Battling burnout.
33. Dealing with loneliness.
34. Struggling with acceptance.
35. Confronting a moral dilemma.
36. Battling a fear of intimacy.
37. Dealing with financial stress.
38. Struggling with faith.
39. Confronting a dark secret.
40. Battling social anxiety.
41. Dealing with a crisis of conscience.
42. Struggling with a loss of purpose.
43. Confronting familial expectations.
44. Battling with a sense of betrayal.
45. Dealing with a fear of death.
46. Struggling with guilt.
47. Confronting a personal weakness.
48. Battling feelings of isolation.
49. Dealing with chronic illness.
50. Struggling with cultural identity.
51. Confronting internalized shame.
52. Battling fear of rejection.
53. Dealing with infertility.
54. Struggling with career dissatisfaction.
55. Confronting past trauma.
56. Battling self-criticism.
57. Dealing with the end of a relationship.
58. Struggling with acceptance from peers.
59. Confronting societal pressures.
60. Battling homesickness.
61. Dealing with unrequited love.
62. Struggling with empathy fatigue.
63. Confronting the fear of the unknown.
64. Battling internal conflict.
65. Dealing with survivor's guilt.
66. Struggling with a feeling of emptiness.
67. Confronting personal hypocrisy.
68. Battling with feelings of insignificance.
69. Dealing with a loss of passion.
70. Struggling with fear of vulnerability.
71. Confronting a fear of judgment.
72. Battling the need for control.
73. Dealing with a fear of change.
74. Struggling with the concept of forgiveness.
75. Confronting feelings of envy.
76. Battling a fear of confrontation.
77. Dealing with feeling unworthy.
78. Struggling with emotional numbness.
79. Confronting a fear of dependency.
80. Battling feelings of helplessness.
81. Dealing with unresolved anger.
82. Struggling with loneliness in a crowd.
83. Confronting the fear of losing loved ones.
84. Battling with feelings of betrayal.
85. Dealing with fear of being forgotten.
86. Struggling with a loss of identity.
87. Confronting the fear of madness.
88. Battling internalized stigma.
89. Dealing with the pressure to succeed.
90. Struggling with a fear of inadequacy.
91. Confronting a fear of public speaking.
92. Battling the need for approval.
93. Dealing with fear of abandonment.
94. Struggling with the concept of mortality.
95. Confronting the need for perfection.
96. Battling feelings of regret.
97. Dealing with a fear of intimacy.
98. Struggling with the fear of failure.
99. Confronting the fear of being alone.
100. Battling internalized anger.

## 4. Moments of Growth: Development and Change

*But life is not all struggle, and so I found solace in* Moments of Growth: Development and Change. *When my character triumphed over obstacles or learned valuable lessons, this oracle guided me in chronicling their evolution. It helped me celebrate small victories and monumental achievements alike, marking the milestones of a life well-lived.*

# 4. Moments of Growth: Development and Change Oracle

1. Learning a new skill.
2. Embracing a new philosophy.
3. Finding a new passion.
4. Letting go of the past.
5. Forgiving someone who hurt you.
6. Accepting a personal truth.
7. Overcoming a major fear.
8. Achieving a long-term goal.
9. Forming a meaningful relationship.
10. Ending a toxic relationship.
11. Standing up for yourself.
12. Setting personal boundaries.
13. Embracing vulnerability.
14. Adopting a healthier lifestyle.
15. Gaining a new perspective.
16. Finding inner peace.
17. Rediscovering a childhood dream.
18. Making a significant life change.
19. Helping someone in need.
20. Taking responsibility for a mistake.
21. Conquering self-doubt.
22. Finding balance in life.
23. Embracing forgiveness.
24. Discovering the power of kindness.
25. Developing a positive self-image.
26. Understanding personal strengths.
27. Cultivating gratitude.
28. Embracing creativity.
29. Strengthening family bonds.
30. Finding community.
31. Pursuing a new career.
32. Embracing independence.
33. Finding a mentor.
34. Becoming a mentor.
35. Volunteering for a cause.
36. Starting a new hobby.
37. Traveling to a new place.
38. Learning from failure.
39. Cultivating resilience.
40. Embracing change.
41. Developing patience.
42. Understanding empathy.
43. Embracing curiosity.
44. Pursuing education.
45. Finding spiritual growth.
46. Overcoming adversity.
47. Learning to let go.
48. Understanding mindfulness.
49. Embracing self-love.
50. Finding joy in simplicity.
51. Understanding the importance of rest.
52. Discovering the value of play.
53. Embracing the power of music.
54. Finding inspiration in art.
55. Developing a daily routine.
56. Embracing the present moment.
57. Understanding the importance of hope.
58. Cultivating optimism.
59. Finding meaning in suffering.
60. Discovering the power of words.
61. Understanding the importance of action.
62. Embracing the unknown.
63. Finding strength in adversity.
64. Understanding the nature of happiness.
65. Cultivating a sense of wonder.
66. Embracing the power of laughter.
67. Finding beauty in imperfection.
68. Understanding the importance of forgiveness.
69. Embracing the journey of self-discovery.
70. Finding solace in nature.
71. Understanding the value of silence.
72. Embracing the spirit of adventure.
73. Finding wisdom in experience.
74. Understanding the power of choice.
75. Embracing the importance of humility.
76. Understanding the significance of trust.
77. Finding peace in acceptance.
78. Understanding the role of intuition.
79. Embracing the concept of destiny.
80. Finding joy in accomplishment.
81. Understanding the impact of love.
82. Cultivating a sense of belonging.
83. Finding clarity in confusion.
84. Understanding the importance of faith.
85. Embracing the journey of healing.
86. Finding inspiration in adversity.
87. Understanding the significance of legacy.
88. Embracing the importance of mindfulness.
89. Finding joy in giving.
90. Understanding the power of resilience.
91. Embracing the concept of harmony.
92. Finding peace in letting go.
93. Understanding the role of courage.
94. Embracing the spirit of generosity.
95. Finding wisdom in ancestors.
96. Understanding the power of dreams.
97. Embracing the value of reflection.
98. Finding meaning in life.
99. Understanding the journey of growth.
100. Embracing the wisdom of the ages.

## 5. Emotional Turmoil: Inner Conflict

*As I delved deeper into my character's emotional landscape,* Emotional Turmoil: Inner Conflict *became an invaluable tool. In moments of inner struggle, when my character's heart was torn between duty and desire, love and loyalty, I consulted this oracle. It guided me through the tempestuous seas of passion and pain, helping me to craft a narrative rich in emotional depth.*

# 5. Emotional Turmoil: Inner Conflict Oracle

1. Conflicted about a recent decision.
2. Torn between two loves.
3. Grappling with guilt over a past action.
4. Struggling with feelings of inadequacy.
5. Battling inner demons.
6. Experiencing a crisis of faith.
7. Dealing with unexpressed anger.
8. Feeling overwhelmed by responsibilities.
9. Struggling with loneliness despite being surrounded by people.
10. Conflicted over a moral dilemma.
11. Dealing with the loss of a loved one.
12. Torn between duty and desire.
13. Struggling with envy towards a friend.
14. Grappling with a secret.
15. Battling fear of the unknown.
16. Experiencing regret over missed opportunities.
17. Struggling to forgive oneself.
18. Conflict about a friendship.
19. Battling feelings of unworthiness.
20. Dealing with shame.
21. Struggling with self-acceptance.
22. Grappling with a fear of intimacy.
23. Battling a sense of betrayal.
24. Feeling torn between tradition and progress.
25. Struggling with conflicting goals.
26. Dealing with conflicting goals.
27. Struggling with a fear of vulnerability.
28. Grappling with jealousy in a relationship.
29. Battling feelings of helplessness.
30. Experiencing a clash of values.
31. Torn between two important people.
32. Struggling with the need for approval.
33. Grappling with a fear of failure.
34. Battling feelings of isolation.
35. Conflicted over a career decision.
36. Dealing with unresolved grief.
37. Struggling with trust issues.
38. Grappling with a fear of change.
39. Battling feelings of resentment.
40. Conflicted about personal beliefs.
41. Struggling with the balance between personal and professional life.
42. Grappling with feelings of inadequacy in a relationship.
43. Battling feelings of unfulfilled potential.
44. Experiencing inner turmoil about the future.
45. Struggling with a sense of identity.
46. Conflicted over family expectations.
47. Grappling with the desire for revenge.
48. Battling feelings of powerlessness.
49. Conflicted about accepting help.
50. Struggling with an ethical dilemma.
51. Grappling with being misunderstood.
52. Battling with societal expectations.
53. Experiencing inner conflict about a past event.
54. Struggling with a fear of being judged.
55. Conflicted about taking a risk.
56. Grappling with the need for control.
57. Battling with feelings of inadequacy at work.
58. Experiencing self-doubt in a relationship.
59. Struggling with the pressure to conform.
60. Conflicted about pursuing a passion.
61. Grappling with feelings of guilt over success.
62. Battling the fear of losing independence.
63. Conflicted about maintaining a secret.
64. Struggling with feelings of inferiority.
65. Grappling with the impact of trauma.
66. Battling inner resistance to change.
67. Experiencing fear of the unknown.
68. Struggling with accepting one's own limitations.
69. Conflicted about a major life decision.
70. Grappling with the fear of being alone.
71. Battling feelings of rejection.
72. Experiencing a clash between personal and societal values.
73. Struggling with the desire to escape reality.
74. Conflicted over a romantic interest.
75. Grappling with the fear of success.
76. Battling feelings of envy towards peers.
77. Experiencing inner conflict about personal growth.
78. Struggling with the impact of external pressure.
79. Conflicted about reconciling with someone.
80. Grappling with feelings of unimportance.
81. Battling the fear of making mistakes.
82. Experiencing doubt about one's path in life.
83. Struggling with a fear of disappointing others.
84. Conflicted about a financial decision.
85. Grappling with feelings of inadequacy as a parent.
86. Battling inner conflict about a personal mission.
87. Experiencing regret over hurting someone.
88. Struggling with the fear of being forgotten.
89. Conflicted about revealing a truth.
90. Grappling with the pressure to succeed.
91. Battling feelings of inner emptiness.
92. Experiencing conflict about moving on.
93. Struggling with the fear of commitment.
94. Conflicted about a significant change.
95. Grappling with feelings of unworthiness of happiness.
96. Battling with feelings of being trapped.
97. Experiencing doubt about one's abilities.
98. Struggling with the fear of taking responsibility.
99. Conflicted about balancing personal needs and others' expectations.
100. Grappling with accepting one's own flaws.

# 6. Important Decisions: Choices That Matter

*The path of a heroine or hero is paved with difficult decisions, and* Decisions That Matter: Choices That Matter *became my compass in those moments. When faced with moral dilemmas or life-changing decisions, I would turn to this oracle for guidance. It challenged me to consider the consequences of my character's actions, adding weight and meaning to each choice.*

# 6. Important Decisions: Choices That Matter Oracle

1. Choosing between two career paths.
2. Deciding whether to trust a stranger.
3. Choosing to forgive someone who wronged you.
4. Deciding to end or mend a relationship.
5. Choosing to reveal a long-held secret.
6. Deciding whether to take a significant risk.
7. Choosing to move to a new place.
8. Deciding to pursue a new hobby or interest.
9. Choosing to stand up for someone.
10. Deciding to confront a fear.
11. Choosing between personal happiness and professional success.
12. Deciding whether to help a former enemy.
13. Choosing to accept or decline a promotion.
14. Deciding to change lifestyle habits.
15. Choosing to start a family.
16. Deciding to reconcile with a family member.
17. Choosing to invest in a business opportunity.
18. Deciding to follow a dream or stay practical.
19. Choosing to seek therapy or counseling.
20. Deciding whether to travel abroad.
21. Choosing to mentor someone.
22. Deciding to forgive yourself for a past mistake.
23. Choosing to adopt a pet.
24. Deciding to volunteer for a cause.
25. Choosing to donate to charity.
26. Deciding to quit a job.
27. Choosing to pursue further education.
28. Deciding whether to attend a significant event.
29. Choosing to buy a home.
30. Deciding to take a sabbatical.
31. Choosing to reconnect with an old friend.
32. Deciding to participate in a community project.
33. Choosing to accept or reject an apology.
34. Deciding whether to relocate for a job.
35. Choosing to start a business.
36. Deciding to write a book or create art.
37. Choosing to take legal action.
38. Deciding to go on an adventure.
39. Choosing to confront someone who hurt you.
40. Deciding whether to keep a promise.
41. Choosing to let go of a grudge.
42. Deciding to make a large purchase.
43. Choosing to change personal appearance.
44. Deciding to make a public speech.
45. Choosing to take a stand on an issue.
46. Deciding to learn a new language.
47. Choosing to adopt a healthier lifestyle.
48. Deciding whether to engage in a protest.
49. Choosing to mentor a younger person.
50. Deciding to plan a significant event.
51. Choosing to join a group or club.
52. Deciding whether to pursue a romantic relationship.
53. Choosing to take on a new responsibility.
54. Deciding to apologize for a wrong.
55. Choosing to forgive a betrayal.
56. Deciding to explore a new culture.
57. Choosing to invest time in a hobby.
58. Deciding whether to move in with someone.
59. Choosing to end a toxic friendship.
60. Deciding to stand up for personal beliefs.
61. Choosing to repair or replace something valuable.
62. Deciding whether to visit a significant place.
63. Choosing to engage in a new sport or activity.
64. Deciding to accept an unexpected opportunity.
65. Choosing to support someone in need.
66. Deciding to take a significant financial risk.
67. Choosing to learn a new skill or trade.
68. Deciding to attend a social event.
69. Choosing to participate in a competition.
70. Deciding to donate time to a cause.
71. Choosing to make a major lifestyle change.
72. Deciding to support a friend or family member.
73. Choosing to participate in a family tradition.
74. Deciding to explore a different religion or spirituality.
75. Choosing to forgive yourself for a mistake.
76. Deciding to trust someone new.
77. Choosing to challenge a personal belief.
78. Deciding whether to change career paths.
79. Choosing to let go of material possessions.
80. Deciding to participate in a cultural event.
81. Choosing to adopt a new mindset.
82. Deciding to make a significant purchase.
83. Choosing to stand up against injustice.
84. Deciding to pursue a creative project.
85. Choosing to confront a fear.
86. Deciding to make amends with someone.
87. Choosing to start or join a support group.
88. Deciding to make a public statement.
89. Choosing to seek reconciliation.
90. Deciding to mentor someone in need.
91. Choosing to take a personal retreat.
92. Deciding to engage in self-care.
93. Choosing to take on a new challenge.
94. Deciding to face a personal flaw.
95. Choosing to learn from a past mistake.
96. Deciding to help a stranger.
97. Choosing to invest in personal development.
98. Deciding to support a cause.
99. Choosing to embrace a new technology.
100. Deciding to create a legacy project.

# 7. Secrets Unveiled: Hidden Truths

*As my tale unfolded,* **Secrets Unveiled: Hidden Truths** *added layers of mystery and revelation to my narrative. This oracle whispered of long-buried secrets, forgotten lore, and hidden agendas. It turned my simple story into an intricate web of truth and deception, keeping me on the edge of my seat as both writer and reader.*

# 7. Secrets Unveiled: Hidden Truths Oracle

1. A family member reveals a hidden history.
2. Discovering a secret about a close friend.
3. Uncovering a hidden talent within yourself.
4. Finding a hidden diary with shocking entries.
5. Learning about a hidden betrayal.
6. Discovering a secret room in a familiar place.
7. Uncovering a hidden message in a letter.
8. Learning about a secret society.
9. Discovering a hidden stash of money.
10. Unveiling a hidden agenda of a trusted ally.
11. Finding out about a secret love affair.
12. Learning about a hidden heirloom.
13. Uncovering a family secret that changes everything.
14. Discovering a hidden weakness of an enemy.
15. Learning about a secret mission.
16. Unveiling a hidden identity.
17. Finding a hidden passageway.
18. Discovering a hidden map.
19. Uncovering a secret document.
20. Learning about a hidden talent of a friend.
21. Discovering a hidden garden.
22. Finding a hidden book with powerful knowledge.
23. Learning about a secret experiment.
24. Uncovering a hidden artifact.
25. Discovering a hidden weapon.
26. Learning about a secret plan.
27. Finding out about a hidden addiction.
28. Unveiling a hidden relationship.
29. Discovering a hidden fear.
30. Learning about a secret alliance.
31. Uncovering a hidden past life.
32. Finding a hidden treasure.
33. Learning about a secret project.
34. Discovering a hidden truth about yourself.
35. Uncovering a hidden motive.
36. Learning about a secret enemy.
37. Discovering a hidden strength.
38. Unveiling a hidden prophecy.
39. Finding a hidden sanctuary.
40. Learning about a secret ritual.
41. Discovering a hidden letter.
42. Unveiling a hidden challenge.
43. Finding out about a secret identity.
44. Learning about a hidden talent within your family.
45. Discovering a hidden family history.
46. Uncovering a secret pact.
47. Learning about a hidden romance.
48. Finding a hidden formula.
49. Discovering a secret connection to a famous person.
50. Unveiling a hidden truth about the world.
51. Learning about a secret prophecy.
52. Discovering a hidden diary.
53. Uncovering a hidden dream.
54. Learning about a secret that changes your perspective.
55. Finding a hidden symbol.
56. Unveiling a hidden curse.
57. Discovering a secret passage in a familiar place.
58. Learning about a hidden agenda.
59. Finding a hidden safe.
60. Uncovering a secret friendship.
61. Discovering a hidden rival.
62. Learning about a secret desire.
63. Finding a hidden relic.
64. Unveiling a hidden betrayal.
65. Discovering a secret about your heritage.
66. Learning about a hidden connection between events.
67. Finding a hidden letter from a loved one.
68. Unveiling a hidden talent.
69. Discovering a secret about a historical event.
70. Learning about a hidden fear within yourself.
71. Uncovering a secret about an old friend.
72. Finding a hidden artifact with mysterious powers.
73. Discovering a hidden love letter.
74. Learning about a secret ritual of an ancient culture.
75. Finding a hidden painting with a mysterious past.
76. Uncovering a hidden conspiracy.
77. Learning about a secret that changes your destiny.
78. Discovering a hidden notebook with important information.
79. Unveiling a hidden portal to another world.
80. Finding a hidden manuscript.
81. Learning about a secret society's plans.
82. Discovering a hidden enemy among your allies.
83. Unveiling a hidden map to a lost city.
84. Finding a hidden message in an old book.
85. Discovering a secret about your mentor.
86. Learning about a hidden resource.
87. Finding a hidden truth in a familiar place.
88. Unveiling a hidden network.
89. Discovering a hidden talent in an unexpected situation.
90. Learning about a secret that changes a relationship.
91. Finding a hidden journal with shocking entries.
92. Uncovering a hidden connection between people.
93. Discovering a hidden trap.
94. Learning about a secret that alters your course.
95. Finding a hidden document with crucial information.
96. Unveiling a hidden truth about your family.
97. Discovering a secret about your past.
98. Learning about a hidden weakness within yourself.
99. Finding a hidden ally in an unlikely place.
100. Uncovering a hidden truth about the world around you.

## 8. Significant Encounters: Influential Meetings

*The oracle of* Significant Encounters: Influential Meetings *became my trusted companion when the world around my character felt stagnant or when the story craved a breath of fresh air. In moments of solitude, when the weight of isolation pressed upon my character's shoulders, I would close my eyes and let my quill dance across the page, invoking this oracle. Suddenly, the tavern door would creak open, revealing a cloaked stranger with eyes that held secrets of distant lands. Or perhaps a letter would arrive, bearing news of a long-lost friend's return. These encounters, born from the oracle's whispers, breathed life into my narrative, creating ripples that would shape the course of my character's destiny.*

## 8. Significant Encounters: Influential Meetings Oracle

1. Meeting a childhood friend after many years.
2. Encountering a wise old mentor.
3. Running into a former rival.
4. Meeting a mysterious stranger.
5. Reuniting with a lost family member.
6. Encountering a famous personality.
7. Meeting a kindred spirit.
8. Running into an old flame.
9. Encountering a traveler with fascinating stories.
10. Meeting a person who changes your perspective.
11. Reuniting with a former teacher.
12. Encountering someone with a shared goal.
13. Meeting an influential leader.
14. Running into an estranged friend.
15. Encountering a person from a different culture.
16. Meeting a future business partner.
17. Reuniting with a former colleague.
18. Encountering someone in desperate need of help.
19. Meeting an artist who inspires you.
20. Running into a person who challenges you.
21. Encountering a healer.
22. Meeting a potential romantic interest.
23. Reuniting with a mentor.
24. Encountering a spiritual guide.
25. Meeting someone who becomes a close friend.
26. Running into a person from a past life.
27. Encountering a rival who has changed.
28. Meeting a stranger who shares a secret.
29. Reuniting with a former adversary.
30. Encountering a wise child.
31. Meeting someone who offers guidance.
32. Running into a person who owes you a favor.
33. Encountering a long-lost relative.
34. Meeting a person with unusual abilities.
35. Reuniting with a past love.
36. Encountering a figure from legend.
37. Meeting a person who influences your career.
38. Running into a person with a shared history.
39. Encountering a wanderer with deep knowledge.
40. Meeting a charismatic leader.
41. Reuniting with an old friend at a significant event.
42. Encountering a person who shares a hidden truth.
43. Meeting someone who becomes a lifelong friend.
44. Running into a former enemy who seeks redemption.
45. Encountering a person who changes your destiny.
46. Meeting a benefactor.
47. Reuniting with a childhood hero.
48. Encountering a person who helps you in a crisis.
49. Meeting a mysterious guide.
50. Running into someone who reveals a hidden talent.
51. Encountering a person who shares your passion.
52. Meeting someone who inspires a new direction.
53. Reuniting with a person from your dreams.
54. Encountering a person who challenges your beliefs.
55. Meeting a stranger who provides a crucial clue.
56. Running into a former ally.
57. Encountering a person who offers forgiveness.
58. Meeting a person who teaches a valuable lesson.

59. Reuniting with someone who offers closure.
60. Encountering a person who helps you find peace.
61. Meeting a fellow traveler on a similar path.
62. Running into a person who rekindles hope.
63. Encountering a mentor who guides your journey.
64. Meeting a person who becomes your confidant.
65. Reuniting with a family member who shares a secret.
66. Encountering a person who reveals your destiny.
67. Meeting a person who shares your vision.
68. Running into someone who helps you in a difficult time.
69. Encountering a person who provides an opportunity.
70. Meeting a person who changes your life course.
71. Reuniting with a former mentor at a crucial moment.
72. Encountering a person who becomes your protector.
73. Meeting a person who reveals a hidden truth.
74. Running into someone who becomes a rival.
75. Encountering a person who offers a new perspective.
76. Meeting a person who brings joy to your life.
77. Reuniting with a friend who has transformed.
78. Encountering a person who shares a prophecy.
79. Meeting someone who influences your future.
80. Running into a person who teaches compassion.
81. Encountering a person who helps you let go of the past.
82. Meeting someone who shares a powerful secret.
83. Reuniting with a person who reveals your strength.
84. Encountering a person who offers redemption.
85. Meeting someone who helps you find your path.
86. Running into a person who helps you heal.
87. Encountering a person who inspires courage.
88. Meeting someone who helps you face your fears.
89. Reuniting with a person who shows you forgiveness.
90. Encountering a person who leads you to a discovery.
91. Meeting someone who becomes your muse.
92. Running into a person who shows you your potential.
93. Encountering a person who shares a crucial message.
94. Meeting someone who helps you find balance.
95. Reuniting with a person who offers wisdom.
96. Encountering a person who teaches humility.
97. Meeting someone who helps you overcome obstacles.
98. Running into a person who reveals your true self.
99. Encountering a person who guides you to success.
100. Meeting someone who shows you the power of love.

## 9. Personal Revelations: Discovering Inner Truths

*When the depths of my character's soul stirred with unspoken truths, I turned to Personal Revelations: Discovering Inner Truths. This oracle served as a mirror, reflecting the hidden aspects of my character's psyche. I found myself using it during quiet moments of introspection – perhaps as my character gazed into the flickering flames of a campfire or stood atop a windswept cliff at dawn. The oracle's gentle prodding would unearth buried memories, forgotten dreams, or repressed emotions. With each revelation, my character's inner landscape transformed, adding layers of complexity to their motivations and actions.*

## 9. Personal Revelations: Discovering Inner Truths Oracle

1. Realizing a hidden talent.
2. Understanding a deep-seated fear.
3. Discovering a forgotten memory.
4. Realizing a long-held belief is false.
5. Understanding the root of a personal issue.
6. Discovering a new passion.
7. Realizing the importance of a relationship.
8. Understanding a recurring dream.
9. Discovering the source of an inner conflict.
10. Realizing a need for change.
11. Understanding a past mistake.
12. Discovering a hidden desire.
13. Realizing your true calling.
14. Understanding a personal strength.
15. Discovering a new purpose.
16. Realizing the impact of a past event.
17. Understanding your own emotions better.
18. Discovering a new perspective on life.
19. Realizing the importance of self-care.
20. Understanding a deep-seated resentment.
21. Discovering a hidden part of your personality.
22. Realizing a need for forgiveness.
23. Understanding your relationship with fear.
24. Discovering a passion for creativity.
25. Realizing the value of honesty.
26. Understanding your own motivations.
27. Discovering a connection to your heritage.
28. Realizing the importance of gratitude.
29. Understanding your own limitations.
30. Discovering a hidden strength.
31. Realizing the value of letting go.
32. Understanding your own desires.
33. Discovering a new outlook on life.
34. Realizing the importance of forgiveness.
35. Understanding your relationship with love.
36. Discovering a new aspect of your identity.
37. Realizing the impact of a decision.
38. Understanding a personal boundary.
39. Discovering the value of patience.
40. Realizing a need for self-improvement.
41. Understanding your relationship with success.
42. Discovering a hidden fear.
43. Realizing the importance of family.
44. Understanding your connection to nature.
45. Discovering a new sense of purpose.
46. Realizing the impact of your actions.
47. Understanding your relationship with failure.
48. Discovering the value of compassion.
49. Realizing a need for personal growth.
50. Understanding your own worth.
51. Discovering a hidden potential.
52. Realizing the value of trust.
53. Understanding your relationship with change.
54. Discovering the importance of self-reflection.
55. Realizing the impact of your words.
56. Understanding your relationship with time.
57. Discovering the value of resilience.
58. Realizing a need for balance.
59. Understanding your connection to others.
60. Discovering a hidden resource.
61. Realizing the importance of boundaries.
62. Understanding your relationship with conflict.
63. Discovering the value of humility.
64. Realizing a need for acceptance.
65. Understanding your relationship with power.
66. Discovering a new appreciation for life.
67. Realizing the importance of inner peace.
68. Understanding your connection to your past.
69. Discovering the value of perseverance.
70. Realizing a need for clarity.
71. Understanding your relationship with joy.
72. Discovering the value of authenticity.
73. Realizing the impact of self-doubt.
74. Understanding your connection to your dreams.
75. Discovering the value of empathy.
76. Realizing a need for self-expression.
77. Understanding your relationship with freedom.
78. Discovering the importance of self-awareness.
79. Realizing the value of dedication.
80. Understanding your connection to your goals.
81. Discovering the impact of your environment.
82. Realizing a need for simplicity.
83. Understanding your relationship with trust.
84. Discovering the value of introspection.
85. Realizing the importance of forgiveness.
86. Understanding your connection to your future.
87. Discovering the value of self-acceptance.
88. Realizing the impact of your choices.
89. Understanding your relationship with courage.
90. Discovering the value of inner strength.
91. Realizing a need for self-respect.
92. Understanding your connection to your community.
93. Discovering the importance of mindfulness.
94. Realizing the value of self-discipline.
95. Understanding your relationship with happiness.
96. Discovering the impact of your thoughts.
97. Realizing a need for personal space.
98. Understanding your connection to your intuition.
99. Discovering the value of inner harmony.
100. Realizing the importance of being present.

## 10. Challenges Faced: Overcoming Obstacles

*The journey of a hero is fraught with trials, and* Challenges Faced: Overcoming Obstacles *became my guide through treacherous waters. When the path ahead seemed insurmountable or when my character's resolve wavered, I would consult this oracle. Its wisdom manifested in myriad forms – a sudden rockslide blocking the mountain pass, a betrayal from a trusted ally, or an internal struggle with fear and self-doubt. These challenges, woven seamlessly into the fabric of my tale, tested my character's mettle and forged their spirit in the crucible of adversity.*

# 10. Challenges Faced: Overcoming Obstacles Oracle

1. Facing a physical challenge.
2. Overcoming a financial obstacle.
3. Dealing with a personal loss.
4. Facing a difficult decision.
5. Overcoming a fear.
6. Dealing with a betrayal.
7. Facing an ethical dilemma.
8. Overcoming self-doubt.
9. Dealing with a difficult relationship.
10. Facing a health issue.
11. Overcoming a professional setback.
12. Dealing with a family conflict.
13. Facing a personal limitation.
14. Overcoming a social obstacle.
15. Dealing with a crisis of faith.
16. Facing a legal issue.
17. Overcoming a creative block.
18. Dealing with a major life change.
19. Facing a moral conflict.
20. Overcoming a mental health challenge.
21. Dealing with a betrayal of trust.
22. Facing a difficult truth.
23. Overcoming an addiction.
24. Dealing with a difficult decision.
25. Facing a career setback.
26. Overcoming a communication barrier.
27. Dealing with a physical limitation.
28. Facing a loss of confidence.
29. Overcoming a misunderstanding.
30. Dealing with a conflict of interest.
31. Facing a cultural challenge.
32. Overcoming an emotional barrier.
33. Dealing with a professional conflict.
34. Facing a fear of the unknown.
35. Overcoming a personal bias.
36. Dealing with a lack of resources.
37. Facing a difficult responsibility.
38. Overcoming a fear of failure.
39. Dealing with a toxic relationship.
40. Facing social injustice.
41. Overcoming a sense of inadequacy.
42. Dealing with a fear of rejection.
43. Facing a difficult transition.
44. Overcoming a limiting belief.
45. Dealing with a fear of change.
46. Facing a moral dilemma.
47. Overcoming a lack of motivation.
48. Dealing with a broken trust.
49. Facing a challenging environment.
50. Overcoming a fear of success.
51. Dealing with a strained relationship.
52. Facing a public challenge.
53. Overcoming a sense of isolation.
54. Dealing with a difficult conversation.
55. Facing a fear of confrontation.
56. Overcoming a lack of direction.
57. Dealing with a professional rivalry.
58. Facing a personal betrayal.
59. Overcoming a fear of the future.
60. Dealing with an emotional trauma.
61. Facing a fear of intimacy.
62. Overcoming a sense of guilt.
63. Dealing with a challenging project.
64. Facing a difficult partnership.
65. Overcoming a lack of confidence.
66. Dealing with a fear of commitment.
67. Facing a challenging deadline.
68. Overcoming a sense of loss.
69. Dealing with a challenging work environment.
70. Facing a difficult decision.
71. Overcoming a fear of the past.
72. Dealing with a fear of the unknown.
73. Facing a personal challenge.
74. Overcoming a sense of helplessness.
75. Dealing with a difficult memory.
76. Facing a fear of being alone.
77. Overcoming a communication breakdown.
78. Dealing with a challenging relationship.
79. Facing a fear of change.
80. Overcoming a lack of support.
81. Dealing with a fear of failure.
82. Facing a fear of rejection.
83. Overcoming a difficult past.
84. Dealing with a challenging situation.
85. Facing a fear of losing control.
86. Overcoming a limiting mindset.
87. Dealing with a difficult task.
88. Facing a fear of uncertainty.
89. Overcoming a challenging situation.
90. Dealing with a fear of the unknown.
91. Facing a personal struggle.
92. Overcoming a sense of inadequacy.
93. Dealing with a difficult environment.
94. Facing a fear of making mistakes.
95. Overcoming a lack of resources.
96. Dealing with a fear of success.
97. Facing a difficult relationship.
98. Overcoming a sense of isolation.
99. Dealing with a fear of the future.
100. Facing a personal challenge.

## 11. Epiphanies: Moments of Clarity

*As my character weathered storms and basked in moments of triumph,* Epiphanies: Moments of Clarity *illuminated the subtle changes in their being. I found myself drawn to this oracle after significant events or decisions, using it to reflect on how these experiences had shaped my character. Perhaps a act of kindness towards an enemy softened a hardened heart, or a brush with mortality ignited a newfound appreciation for life's simple pleasures. These moments of growth, sometimes grand and sometimes whisper-quiet, added depth and authenticity to my character's evolution.*

# 11. Epiphanies: Moments of Clarity Oracle

1. The true meaning behind a recurring dream is suddenly revealed.
2. A long-held belief is shattered, opening new perspectives.
3. The solution to a seemingly impossible problem becomes clear.
4. A pattern in past events emerges, revealing a hidden truth.
5. The real motivation behind a rival's actions is understood.
6. A misinterpreted memory is seen in a new light.
7. The consequences of a past decision fully dawn on you.
8. A prophecy's true meaning becomes apparent.
9. The perfect words to express a complex emotion are found.
10. A childhood fear is understood and overcome.
11. The key to mastering a challenging skill is grasped.
12. A loved one's actions are seen from a new perspective.
13. The interconnectedness of all things is momentarily perceived.
14. A cryptic message from the past suddenly makes sense.
15. The root of a long-standing personal flaw is realized.
16. A hidden talent or ability is discovered within oneself.
17. The true nature of a magical artifact is comprehended.
18. A mysterious symbol's meaning is deciphered.
19. The perfect strategy to overcome a formidable foe is conceived.
20. A life-changing decision's ramifications become clear.
21. The beauty in a seemingly mundane object is truly seen.
22. A complex philosophical concept is suddenly understood.
23. The source of inner strength is recognized.
24. A betrayal's deeper implications are fully grasped.
25. The key to reconciling with an estranged friend is realized.
26. A long-forgotten promise's importance is remembered.
27. The true value of a seemingly worthless object is recognized.
28. A recurring nightmare's message is deciphered.
29. The perfect solution to a moral dilemma presents itself.
30. A character flaw's origin in past trauma is understood.
31. The means to break a curse becomes apparent.
32. A cryptic prophecy's double meaning is unraveled.
33. The key to unlocking a hidden power is discovered.
34. A perceived enemy's true intentions are revealed.
35. The perfect words to inspire others are found.
36. A painful memory's lesson is finally learned.
37. The true nature of a mysterious illness is understood.
38. A complex political situation's core issue becomes clear.
39. The perfect gift for a difficult-to-please person is realized.
40. A long-standing grudge's futility is recognized.
41. The key to achieving a long-held dream is grasped.
42. A misunderstood allies' loyalty is fully appreciated.
43. The solution to a logistical nightmare presents itself.
44. A mysterious location's significance is comprehended.
45. The perfect way to honor a fallen comrade is conceived.
46. A seemingly random series of events' purpose is understood.
47. The key to mastering a forbidden technique is realized.
48. A perceived weakness' potential as a strength is seen.
49. The perfect argument to sway an opponent is formulated.

50. A long-lost artifact's location suddenly becomes obvious.
51. The true meaning of an ancient text is deciphered.
52. A character-defining moment's full impact is realized.
53. The key to breaking a destructive habit is discovered.
54. A rival's hidden vulnerability is perceived.
55. The perfect balance between two conflicting needs is found.
56. A mysterious disappearance's cause becomes apparent.
57. The key to uniting warring factions is conceived.
58. A trusted mentor's flaws are recognized and accepted.
59. The perfect way to convey a difficult truth is realized.
60. A seemingly insurmountable obstacle's weakness is found.
61. The true cost of achieving a goal is fully understood.
62. A recurring conflict's underlying cause is recognized.
63. The key to unlocking a sealed memory is discovered.
64. A complex magical theory is suddenly comprehended.
65. The perfect response to a long-standing insult is conceived.
66. A missed opportunity's silver lining becomes apparent.
67. The key to breaking a magical illusion is grasped.
68. A trusted institution's corruption is fully realized.
69. The perfect way to honor one's heritage is understood.
70. A long-standing fear's irrationality is recognized.
71. The key to decoding a secret language is discovered.
72. A seemingly random collection's true value is perceived.
73. The perfect method to train a difficult skill is conceived.
74. A long-held secret's burden is fully understood.
75. The key to resolving inner conflict is realized.
76. A misunderstood prophecy's true warning is grasped.
77. The perfect time to reveal a hidden truth presents itself.
78. A complex relationship's underlying dynamic is understood.
79. The key to unlocking a sealed door is discovered.
80. A perceived failure's value as a learning experience is seen.
81. The perfect way to make amends for past wrongs is conceived.
82. A long-standing mystery's solution suddenly becomes clear.
83. The key to mastering one's emotions is realized.
84. A trusted weapon's hidden potential is discovered.
85. The perfect words to break a magical enchantment are found.
86. A recurring dream's connection to reality is understood.
87. The key to overcoming a legendary challenge is grasped.
88. A perceived enemy's similar goals are recognized.
89. The perfect way to preserve a dying tradition is conceived.
90. A long-forgotten skill's relevance becomes apparent.
91. The key to unlocking inner peace is discovered.
92. A complex puzzle's solution suddenly becomes obvious.
93. The perfect balance between duty and desire is found.
94. A trusted ally's hidden agenda is fully comprehended.
95. The key to breaking a generational curse is realized.
96. A misinterpreted sign's true meaning is understood.
97. The perfect way to honor conflicting vows is conceived.
98. A long-standing rivalry's pointlessness is recognized.
99. The key to transcending mortal limitations is grasped.
100. The true nature of one's destiny is fully understood.

## 12. Reflective Moments: Introspection and Contemplation

*In the stillness between adventures, when the ink had dried on tales of daring deeds, I turned to* Reflective Moments: Introspection and Contemplation. *This oracle guided my quill during moments of quiet reflection – a solitary walk through an autumn forest, a sleepless night beneath unfamiliar stars, or the calm before a looming storm. It prompted my character to ponder the weight of their choices, the nature of their existence, and the legacy they wished to leave behind. These introspective passages became the heartbeat of my journal, grounding the epic tale in the rich soil of human experience.*

## 12. Reflective Moments: Introspection and Contemplation Oracle

1. Reflecting on a significant life event.
2. Contemplating a recent decision.
3. Reflecting on a meaningful relationship.
4. Contemplating your purpose in life.
5. Reflecting on personal achievements.
6. Contemplating your values and beliefs.
7. Reflecting on a life lesson learned.
8. Contemplating a recent challenge.
9. Reflecting on a personal growth moment.
10. Contemplating your dreams and aspirations.
11. Reflecting on a past mistake.
12. Contemplating your fears and anxieties.
13. Reflecting on your childhood.
14. Contemplating your future.
15. Reflecting on a significant change in your life.
16. Contemplating your relationships with others.
17. Reflecting on a personal transformation.
18. Contemplating your sense of identity.
19. Reflecting on your connection to nature.
20. Contemplating the meaning of success.
21. Reflecting on your life journey.
22. Contemplating your spiritual beliefs.
23. Reflecting on a memorable experience.
24. Contemplating your role in your community.
25. Reflecting on your personal strengths.
26. Contemplating your weaknesses.
27. Reflecting on your happiest moments.
28. Contemplating your greatest fears.
29. Reflecting on a recent conversation.
30. Contemplating your goals and ambitions.
31. Reflecting on your daily routines.
32. Contemplating your self-care practices.
33. Reflecting on a book or film that influenced you.
34. Contemplating your creative expression.
35. Reflecting on your work or career.
36. Contemplating your financial goals.
37. Reflecting on your health and wellness.
38. Contemplating your friendships.
39. Reflecting on your family dynamics.
40. Contemplating your personal boundaries.
41. Reflecting on your educational experiences.
42. Contemplating your travel experiences.
43. Reflecting on a moment of kindness.
44. Contemplating your regrets.
45. Reflecting on your personal inspirations.
46. Contemplating your connection to your heritage.
47. Reflecting on a dream or aspiration.
48. Contemplating your emotional well-being.
49. Reflecting on your habits and behaviors.
50. Contemplating your life balance.
51. Reflecting on a meaningful conversation.
52. Contemplating your spiritual journey.
53. Reflecting on your cultural influences.
54. Contemplating your legacy.
55. Reflecting on your personal philosophy.
56. Contemplating your favorite memories.
57. Reflecting on your lifestyle choices.
58. Contemplating your personal challenges.
59. Reflecting on your milestones.
60. Contemplating your sense of adventure.
61. Reflecting on a turning point in your life.
62. Contemplating your resilience.
63. Reflecting on your self-worth.
64. Contemplating your empathy towards others.
65. Reflecting on your sense of humor.
66. Contemplating your adaptability.
67. Reflecting on your achievements and failures.
68. Contemplating your emotional intelligence.
69. Reflecting on your moments of bravery.
70. Contemplating your forgiveness journey.
71. Reflecting on your decision-making process.
72. Contemplating your personal rituals.
73. Reflecting on your sense of gratitude.
74. Contemplating your creativity.
75. Reflecting on your impact on others.
76. Contemplating your relationship with time.
77. Reflecting on your personal sacrifices.
78. Contemplating your sources of joy.
79. Reflecting on your personal transformations.
80. Contemplating your life's purpose.
81. Reflecting on your dreams and goals.
82. Contemplating your personal growth.
83. Reflecting on your inner peace.
84. Contemplating your life's lessons.
85. Reflecting on your support systems.
86. Contemplating your resilience in adversity.
87. Reflecting on your aspirations for the future.
88. Contemplating your life experiences.
89. Reflecting on your inner conflicts.
90. Contemplating your personal strengths.
91. Reflecting on your emotional growth.
92. Contemplating your relationships with loved ones.
93. Reflecting on your moments of clarity.
94. Contemplating your personal values.
95. Reflecting on your spiritual connections.
96. Contemplating your life's turning points.
97. Reflecting on your personal milestones.
98. Contemplating your sources of inspiration.
99. Reflecting on your sense of fulfillment.
100. Contemplating your journey towards self-discovery.

# LAST WORDS

There exists an arcane tradition known as Solo Journaling RPGs. This ancient art, once guarded by the enigmatic Oracle Keepers, has been passed down through generations of dreamers, knights, and wandering souls. It is said that Siwa, the first Oracle, discovered the power to transcend reality through her writings, her quill opening portals to realms unseen.

Solo Journaling RPGs are a window into the depths of one's own imagination, a journey of introspection and personal storytelling. Here, the player becomes both the protagonist and the chronicler, weaving tales of triumph, heartache, and self-discovery. With each stroke of the pen, characters come alive, their thoughts and emotions spilling onto the page in vivid detail.

At the heart of this mystical practice lie the oracles - tools of divination that guide the narrative and spark creativity. These oracles are more than mere tables or lists; they are conduits to the arcane tapestry of storytelling. They whisper of personal revelations, significant encounters, and moments of profound growth. They unveil secrets long buried and illuminate paths yet untrodden.

To begin a solo journaling RPG is to dance with fate itself. As you write, you may find yourself unearthing memories from your character's past, confronting their deepest fears, or celebrating their most cherished victories. The oracles will lead you through labyrinths of emotion, challenge you with moral dilemmas, and inspire you with visions of possible futures.

Remember that your journal is a living artifact. Each entry weaves a new thread in the grand tapestry of your character's life. The Oracle Keepers may have vanished, but their legacy lives on in every word you inscribe. Your quill is your wand, your imagination the source of magic.

You stand at the threshold of infinite possibilities. The oracles await your touch, ready to guide you through personal struggles, moments of clarity, and the complex maze of human emotion. They will help you explore the legacy of your character's past and the conflicting desires that shape their present.

In this solitary yet profoundly connected experience, you'll find that playing alone is never truly lonely. The worlds you create, the characters you encounter, and the journeys you undertake will resonate with the echoes of countless storytellers who came before you.

So take a deep breath, open your journal, and let the ink flow. Trust in the oracles, but remember that the true power lies within your own imagination. Your story awaits, a tale yet untold, a journey that begins with a single word.

In the preceding pages, you have a collection of oracles, each designed to enrich your personal narrative and create a journal full of life and detail. These are your tools, your guides, your whispers of inspiration. Use them wisely, let them challenge you, and allow them to lead you to unexpected revelations.

Now, the stage is set. The oracles stand ready. Your quill hovers above the page. What tales will you tell? What worlds will you explore? The journey of a thousand stories begins with a single word. Are you ready to write your legend?

# SOLO RPG JOURNALING

## THE ART OF BEING YOUR OWN STORYTELLER

BASUNAT

Made in the USA
Las Vegas, NV
01 November 2024

10911190R00020